Use of this workbook is limited to educational purposes.

This workbook may not be copied, sold, or resold from any person, entity, or organization other than the originating designer, John A. Honeycutt.

All content, images, and thought-leadership are owned by their respective copyright holder and subject to those rights in addition to restrictions imposed by John A. Honeycutt.

Processes associated with design and layout of curriculum is patent-pending by John A. Honeycutt as Honeycutt 21st Century Instructional Design. Design, layout, and content not otherwise owed by other entities is owned by John A. Honeycutt.

HoneycuttScience Work Book

CHEMISTRY

CH_4 + $2O_2$ ⟶ CO_2 + $2H_2O$

Copyright John A. Honeycutt 2017. All rights reserved.

Contents

11.1	Welcome to Chemistry	1
12.1	Nature of Matter	7
13.1	Properties of Matter	13
14.1	The Elements	19
15.1	Using the Periodic Table	25
16.1	Naming Binary Compounds	31
17.1	Naming and Writing Formulas	37
18.1	Scientific Notation and Units	43
19.1	Atoms and Moles	49
23.1	Reactions in Aqueous Solutions	55
	GLOSSARY	61

11.1 Introduction to Chemistry

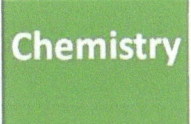

Summarize main points from each video.

Video Title / topic

Video Title / topic

Video Title / topic

Topic Introduction

Summarize your understanding of each paragraph.

All materials on Earth are composed of different combinations of atoms. Atoms are the smallest particles of a chemical element that still have all the unique chemical properties of this element.

Atoms and compounds they form play a role in almost all processes occurring on Earth and in space. All living organisms rely on a set of chemical compounds and chemicals to digest food, energy transport reactions, and to reproduce.

Protons are positively charged (+1), quite the opposite, as the electric charge of electrons (-1). The number of protons in the nucleus determines the total amount of positive charge in the atom.

In electrically neutral atom, the number of protons and electrons is equal, such that positive and negative charges are balanced to zero. Neutrons are about the same size as protons, but slightly heavier.

http://www.ency123.com/2014/03/protons-neutrons-and-electrons.html

Read/Summarize Text

1. **Read the passage.**
2. **Underline key expressions in each sentence.**
3. **Re-write each word (or expression) you underlined.**
4. **Summarize the passage.**

Title of Passage.

Definition of a Scientific Law. A scientific law is a statement that describes an observable occurrence in nature that appears to always be true. It is a term used in all of the natural sciences (astronomy, biology, chemistry and physics, to name a few).

Definition of atom. The smallest component of an element having the chemical properties of the element, consisting of a nucleus containing combinations of neutrons and protons and one or more electrons bound to the nucleus by electrical attraction; the number of protons determines the identity of the element

http://www.dictionary.com/browse/atom

Re-write words you underlined

_____   _____   _____

_____   _____   _____

Using a complete sentence, summarize or rephrase the passage

3

Read Text for Comprehension

Read this article for deeper understanding. No summary is required, although you may want to circle, underline, or mark key ideas and words.

Wikibooks has a weblink for High School Chemistry Students. The topic index includes these subject areas:

The Science of Chemistry; Chemistry - A Physical Science; Chemistry in the Laboratory; The Atomic Theory; The Bohr Model of the Atom; Quantum Mechanics Model of the Atom; Electron Configurations for Atoms; Electron Configurations and the Periodic Table; Relationships Between the Elements; Trends on the Periodic Table; Ions and the Compounds They Form; Writing and Naming Ionic Formulas; Covalent Bonding; Molecular Architecture; The Mathematics of Compounds; Chemical Reactions; Mathematics and Chemical Equations; The Kinetic-Molecular Theory; The Liquid State; The Solid State; The Solution Process; Ions in Solution; Chemical Kinetics; Chemical Equilibrium; Acids; Water, pH, and Titration; Thermodynamics; Electrochemistry; Nuclear Chemistry; Organic Chemistry.

The article and associated information is found at this link:

https://en.wikibooks.org/wiki/High_School_Chemistry

Additionally, you may download a free 197 page pdf-format workbook found at:

https://upload.wikimedia.org/wikipedia/commons/3/37/High_School_Chemistry_Workbook.pdf

While students are not required to have internet access at home to complete this course, this suggested free resources is both informative and convenient for those with web access.

https://en.wikibooks.org

Draw Illustration

Copy and Label the Illustration in the Space Provided

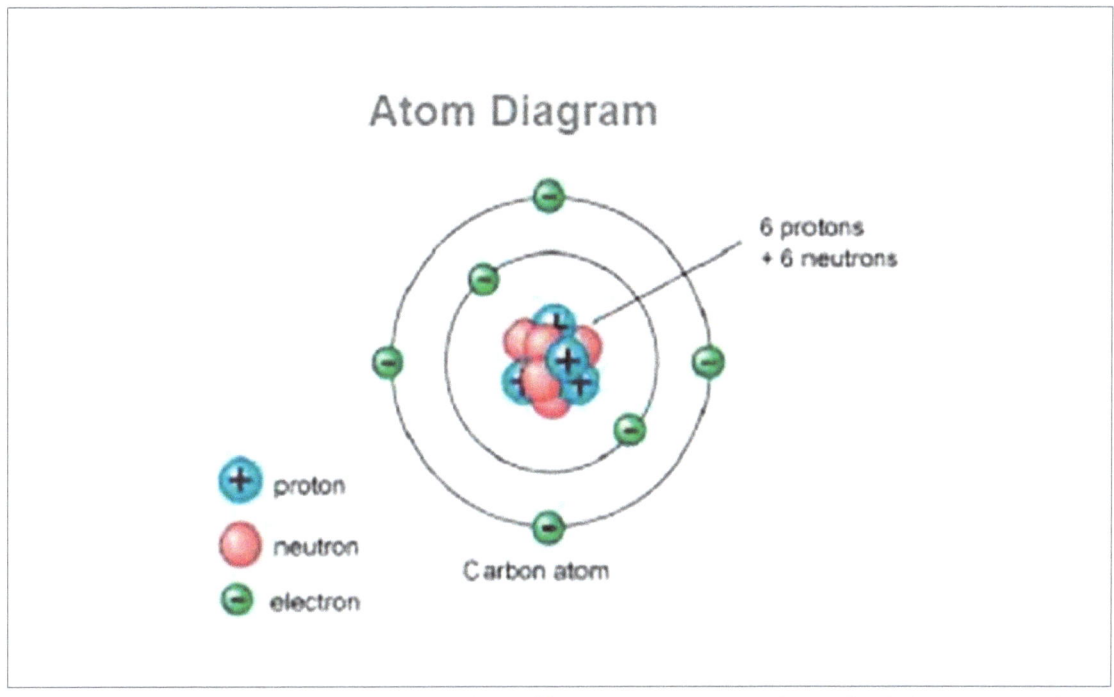

http://www.ency123.com/2013/07/what-is-atom.html

Draw (Copy) the Illustration Here

Interpret a Graph

Write the title of the graph _____

Circle the type of chart this represents

    Bar Chart   Line Chart   Pie Chart   Other

If applicable,

    What does the X-axis represent _____

    What does the Y-axis imply _____

Summarize what this graph represents or conveys

_____

www.compoundchem.com

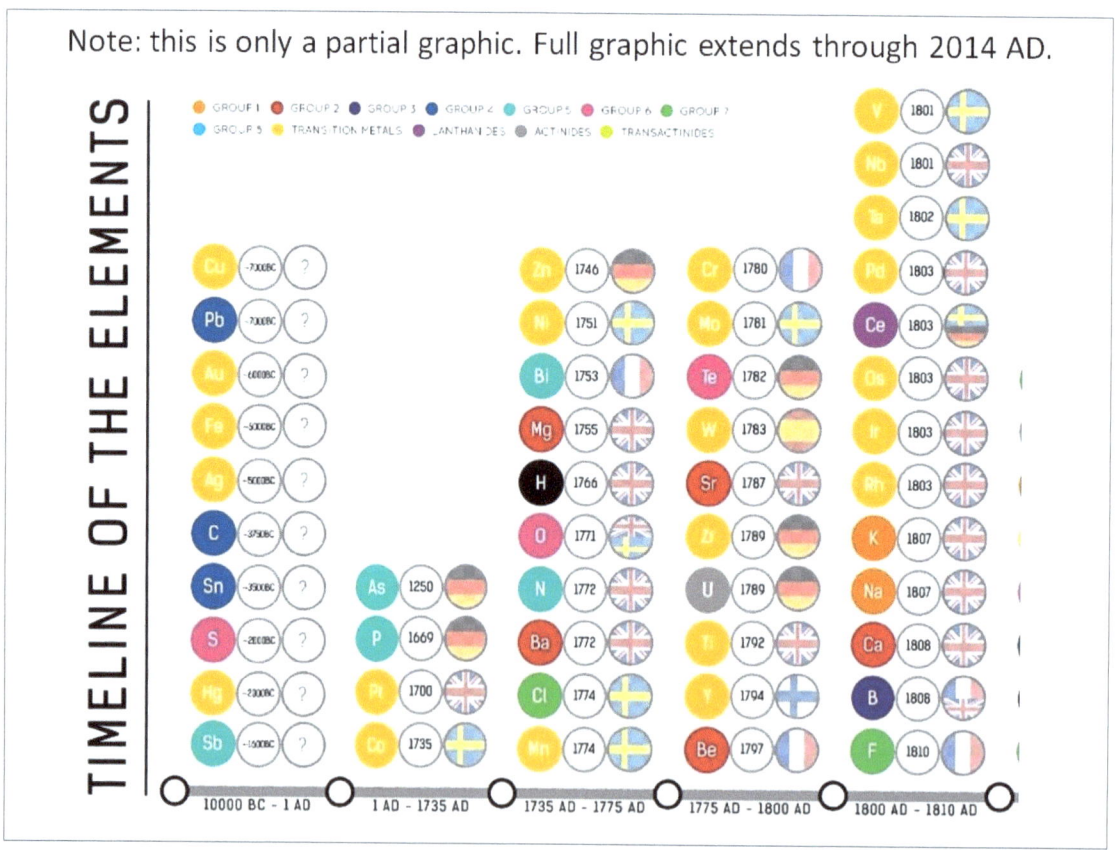

12.1 The Nature of Matter

Summarize main points from each video.

Video Title / topic _____

Video Title / topic _____

Video Title / topic _____

Topic Introduction

Summarize your understanding of each paragraph.

Matter is the substance of which all material is made. That means objects which have mass. Ordinary matter is made of tiny particles called atoms.

[ ]

Matter is the **Stuff Around You**. Matter is everything around you. Atoms and molecules are all composed of matter. Matter is anything that has mass and takes up space.

[ ]

While "matter" has many definitions, a common definition is that it is any substance which has mass and occupies space. All physical objects are composed of matter, in the form of atoms, which are in turn composed of protons, neutrons, and electrons.

[ ]

According to modern physics, matter consists of various types of particles, each with mass and size. The most familiar examples of material particles are the electron, the proton and the neutron. ... Matter can exist in several states, also called phases.

[ ]

Read/Summarize Text

1. **Read the passage.**
2. **Underline key expressions in each sentence.**
3. **Re-write each word (or expression) you underlined.**
4. **Summarize the passage.**

What is mass?

In physics, mass is a property of a physical body. It is the measure of an object's resistance to acceleration (a change in its state of motion) when a net force is applied. It also determines the strength of its mutual gravitational attraction to other bodies. The basic SI unit of mass is the kilogram (kg). Mass is not the same as weight, even though mass is often determined by measuring the object's weight using a spring scale, rather than comparing it directly with known masses. An object on the Moon would weigh less than it does on Earth because of the lower gravity, but it would still have the same mass. Weight is a force, while mass is the property that (along with gravity) determines the strength of this force.

https://en.wikipedia.org/wiki/Mass

Re-write words you underlined

_____   _____   _____

_____   _____   _____

Using a complete sentence, summarize or rephrase the passage

Read Text for Comprehension

Read this article for deeper understanding. No summary is required, although you may want to circle, underline, or mark key ideas and words.

About SI ...

The International System of Units (abbreviated SI from systeme internationale, the French version of the name) is a scientific method of expressing the magnitudes or quantities of important natural phenomena. There are seven base units in the system, from which other units are derived.

The ampere (A)
The ampere is the SI base unit of electrical current.

The candela (cd)
The candela is the SI base unit of luminous intensity.

The kelvin (K)
The kelvin is the SI base unit of thermodynamic temperature.

The kilogram (kg)
The kilogram is the SI base unit of mass.

The metre (m)
The metre is the SI base unit of length.

The mole (mol)
The mole is the SI base unit of amount of substance.

The second (s)
The second is the SI base unit of time.

http://www.npl.co.uk

Draw Illustration

Copy and Label the Illustration in the Space Provided

WHAT IS MATTER?

Matter: A substance that has mass and volume (takes up space).

http://slideplayer.com

Draw (Copy) the Illustration Here

Interpret a Graph

Write the title of the graph _____

Circle the type of chart this represents
        Bar Chart   Line Chart   Pie Chart   Other

If applicable,
        What does the X-axis represent _____

        What does the Y-axis imply _____

Summarize what this graph represents or conveys

_____

http://www.madsci.org

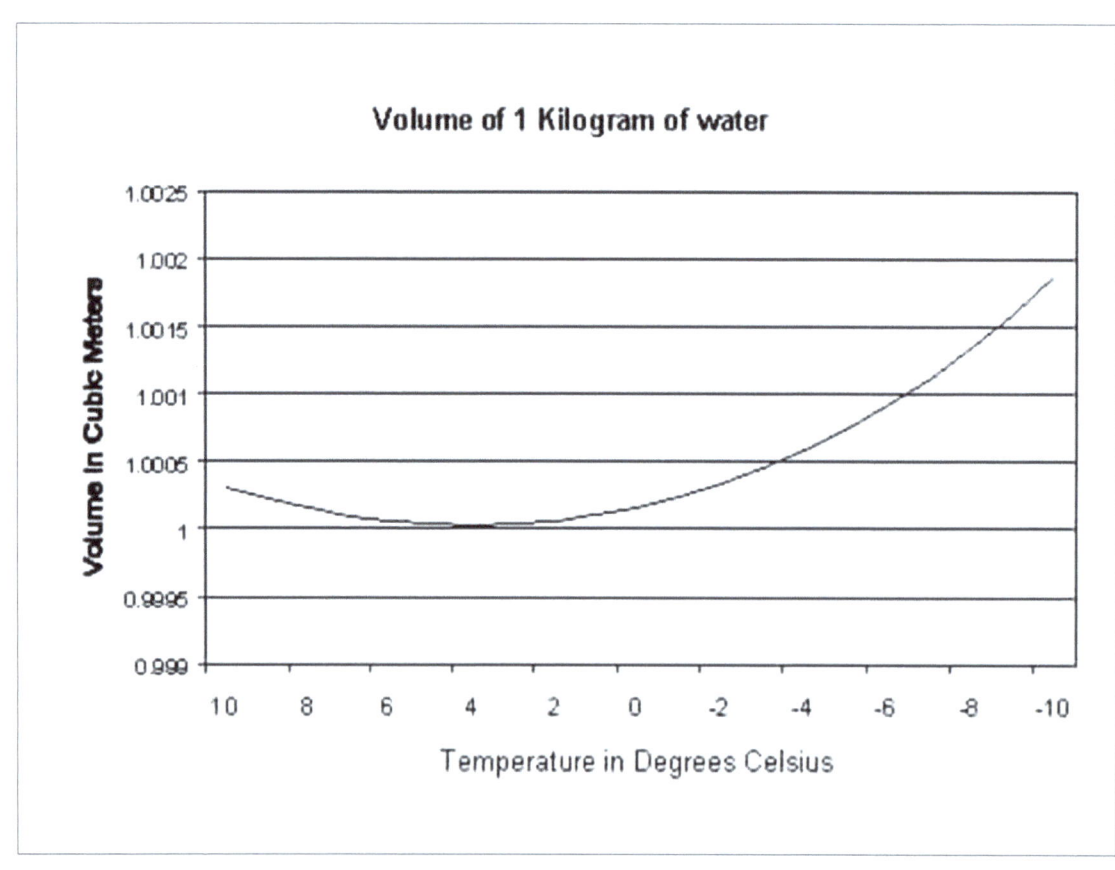

13.1 Properties of Matter

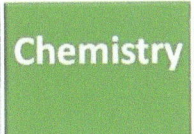

Summarize main points from each video.

Video Title / topic

Video Title / topic

Video Title / topic

Topic Introduction

Summarize your understanding of each paragraph.

Often (typically) introductory courses to chemistry focus on "inorganic" chemistry. Even so, principles of inorganic chemistry generally ALSO apply to organic chemistry. Inorganic compounds might contain hydrogen or carbon, but if they have both, they are organic.

Organic chemistry is the study of molecules that contain carbon compounds. In contrast, inorganic chemistry is the study of all compounds that do NOT contain carbon compounds.

In chemistry terms, organic means that a molecule has a "carbon backbone" with "some hydrogens thrown in" for good measure. Living creatures are made of various kinds of organic compounds. Inorganic molecules are composed of other elements.

The molecule CO_2 (carbon dioxide) is generally considered inorganic. CO_2 at high pressure and temperature is used to remove the organic caffeine molecule from coffee. Read through the longer article in this packet for information about this process.

Read/Summarize Text

1. Read the passage.
2. Underline key expressions in each sentence.
3. Re-write each word (or expression) you underlined.
4. Summarize the passage.

Title of Passage. _____

In thermodynamics, the triple point of a substance is the temperature and pressure at which the three phases (gas, liquid, and solid) of that substance coexist in thermodynamic equilibrium.

0.01 °C & 612 pascals. The single combination of pressure and temperature at which liquid water, solid ice, and water vapor can coexist in a stable equilibrium occurs at exactly 273.16 K (0.01 °C; 32.02 °F) and a partial vapor pressure of 611.657 pascals (6.11657 mbar; 0.00603659 atm). At that point, it is possible to change all of the substance to ice, water, or vapor by making arbitrarily small changes in pressure and temperature.

https://en.wikipedia.org/wiki/Triple_point

Re-write words you underlined

_____ _____ _____

_____ _____ _____

Using a complete sentence, summarize or rephrase the passage

Read Text for Comprehension

Read this article for deeper understanding. No summary is required, although you may want to circle, underline, or mark key ideas and words.

A chemical compound is termed inorganic if it fulfills one or more of the following criteria:

- There is an absence of carbon in its composition
- It is of a non-biologic origin
- It cannot be found or incorporated into a living organism

There is no clear or universally agreed-upon distinction between organic and inorganic compounds. Organic chemists traditionally and generally refer to any molecule containing carbon as an organic compound and by default this means that inorganic chemistry deals with molecules lacking carbon

As many minerals are of biological origin, biologists may distinguish organic from inorganic compounds in a different way that does not hinge on the presence of a carbon atom. Pools of organic matter, for example, that have been metabolically incorporated into living tissues persist in decomposing tissues, but as molecules become oxidized into the open environment, such as atmospheric CO_2, this creates a separate pool of inorganic compounds

---

Caffeine – an organic molecule – is removed from coffee beans through use an inorganic molecule - carbon dioxide (CO_2) – under high pressure.

Supercritical fluid extraction using carbon dioxide is now being widely used as a more effective and environmentally friendly decaffeination method. At temperatures above 304.2 K and pressures above 7376 kPa, CO_2 is a supercritical fluid, with properties of both gas and liquid. Like a gas, it penetrates deep into the coffee beans; like a liquid, it effectively dissolves certain substances. Supercritical carbon dioxide extraction of steamed coffee beans removes 97–99% of the caffeine, leaving coffee's flavor and aroma compounds intact.

Because CO_2 is a gas under standard conditions, its removal from the extracted coffee beans is easily accomplished, as is the recovery of the caffeine from the extract. The caffeine recovered from coffee beans via this process is a valuable product that can be used subsequently as an additive to other foods or drugs.

https://en.wikipedia.org/wiki/Inorganic_compound
https://courses.lumenlearning.com

Draw Illustration

Copy and Label the Illustration in the Space Provided

Illustration

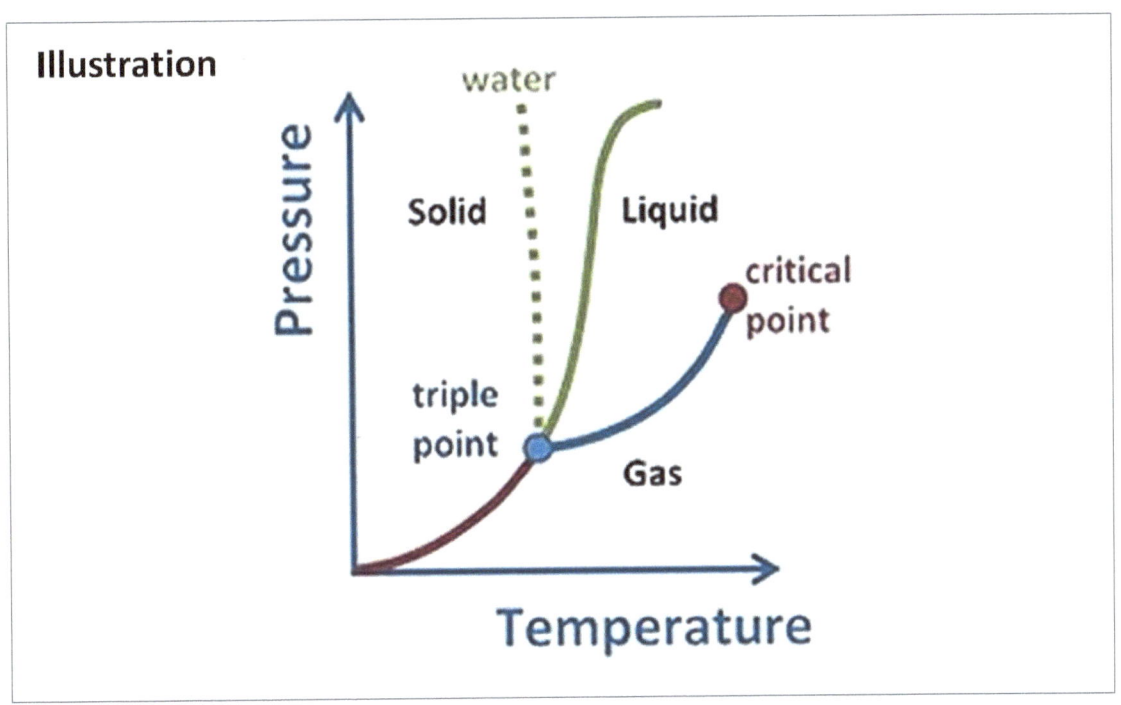

https://en.wikipedia.org/wiki/Matter

Draw (Copy) the Illustration Here

Interpret a Graph

Write the title of the graph _____

Circle the type of chart this represents

    Bar Chart   Line Chart   Pie Chart   Other

If applicable,

    What does the X-axis represent _____

    What does the Y-axis imply _____

Summarize what this graph represents or conveys

_____

https://en.wikipedia.org/wiki/Volatility_(chemistry)

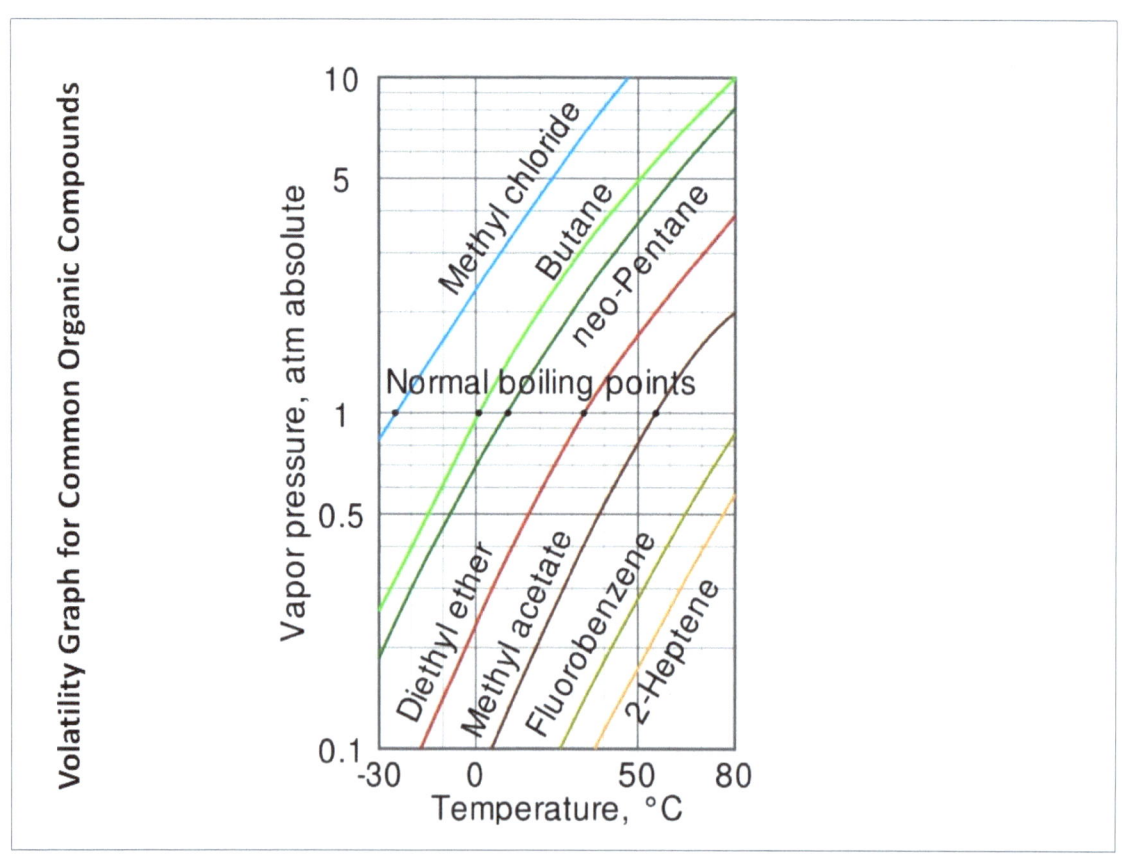

14.1 The Elements

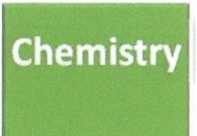

Summarize main points from each video.

Video Title / topic

Video Title / topic

Video Title / topic

Topic Introduction

Summarize your understanding of each paragraph.

Element(s) often refers to the elements of chemistry, each a pure substance of one type of atom, which together make up all the matter in the universe. The periodic table of elements displays all of the elements and their defining attributes.

A chemical element is a species of atoms having the same number of protons in their atomic nuclei (i.e. the same atomic number, or Z). There are 118 elements that have been identified, of which the first 94 occur naturally on Earth. The remaining 24 are synthetic elements.

When different elements are chemically combined, with the atoms held together by chemical bonds, they form chemical compounds. Only a minority of elements are found uncombined as relatively pure minerals.

Common native elements are copper, silver, gold, carbon (as coal, graphite, or diamonds), and sulfur. All but a few of the most inert elements, such as noble gases and noble metals, are usually found on Earth in chemically combined form, as chemical compounds.

Read/Summarize Text

1. Read the passage.
2. Underline key expressions in each sentence.
3. Re-write each word (or expression) you underlined.
4. Summarize the passage.

Where can I find pure elements around my house?

Q. My chemistry teacher is having us put together an "Element Collection" and I want to be able to make a good one. I've already got carbon from burning sugar, and my teacher said that aluminum foil is pure aluminum. He also said gold-plated or silver plated objects count as gold or silver. But he said that you can't bring him a cup of water and say "Here's pure hydrogen mixed with pure oxygen." Any suggestions of easy to find pure elements?

A. A piece of iron, not steel; A neon lamp; A piece of copper; A piece of zinc; A thermometer containing mercury; A piece of lead (the metal, not graphite from a pencil).

https://answers.yahoo.com

Re-write words you underlined

_____ _____ _____

_____ _____ _____

Using a complete sentence, summarize or rephrase the passage

21

Read Text for Comprehension

Read this article for deeper understanding. No summary is required, although you may want to circle, underline, or mark key ideas and words.

ACI ALLOYS, INC

ACI Alloys works with all non-radioactive metals in the periodic table, as well as their alloys and ceramics. Due to the nearly infinite number of possible combinations, we include here only the MSDS sheets for the pure elements. For questions involving properties of alloys or ceramics, please contact us.

ACI ALLOYS understands the importance of quality for both R&D and production materials. We have developed an in-house process to ensure that high quality, reproducible products are made for even the most unusual alloy combinations.

Here are some recent rare-earth sputtering targets we have made:

- Cerium and cerium alloys (cerium-gadolidium, cerium-samarium)
- Dysprosium and dysprosium alloys (iron-dysprosium-terbium)
- Erbium and erbium alloys (gadolinium-erbium-silicon)
- Europium and europium alloys (aluminum-barium-europium, barium-europium)
- Gadolinium and gadolidium alloys (cobalt-gadolinium, iron-gadolinium, gadolinium-terbium)
- Holmium and holmium alloys (holmium-zirconium)
- Lanthanum and lanthanum alloys (lanthanum-nickel)
- Lutetium and lutetium alloys (gold-lutetium, silver-lutetium, tin-silver-lutetium)
- Neodymium and neodymium alloys (neodymium-iron-boron, aluminum-neodymium)
- Praeseodymium and praeseodymium alloys (praeseodymium-iron-boron)
- Samarium and samarium alloys (samarium-cobalt, silver-samarium, samarium-iron)
- Scandium and scandium alloys (aluminum-scandium, chromium-scandium, nickel-scandium, scandium-titanium)

http://www.acialloys.com

Draw Illustration

Copy and Label the Illustration in the Space Provided

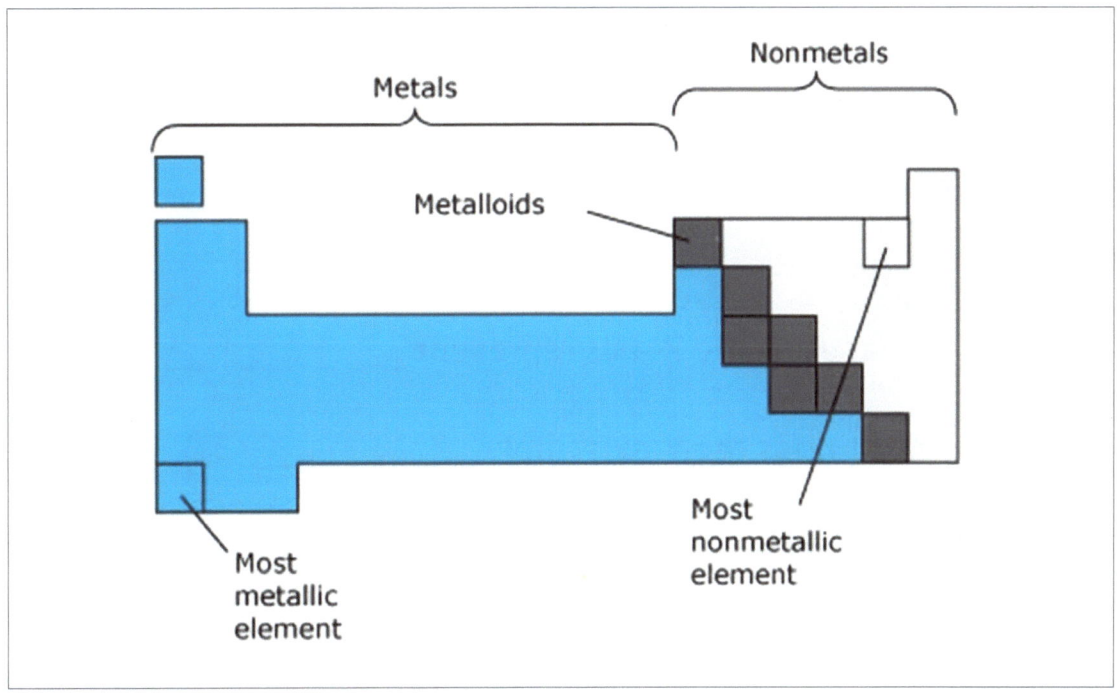

© 2007 – 2009 The University of Waikato | www.sciencelearn.org.nz   www.sciencelearn.org.nz

Draw (Copy) the Illustration Here

23

Interpret a Graph

Write the title of the graph _____

Circle the type of chart this represents

      Bar Chart   Line Chart   Pie Chart   Other

If applicable,

    What does the X-axis represent _____

    What does the Y-axis imply _____

Summarize what this graph represents or conveys

_____

http://www.businessinsider.com

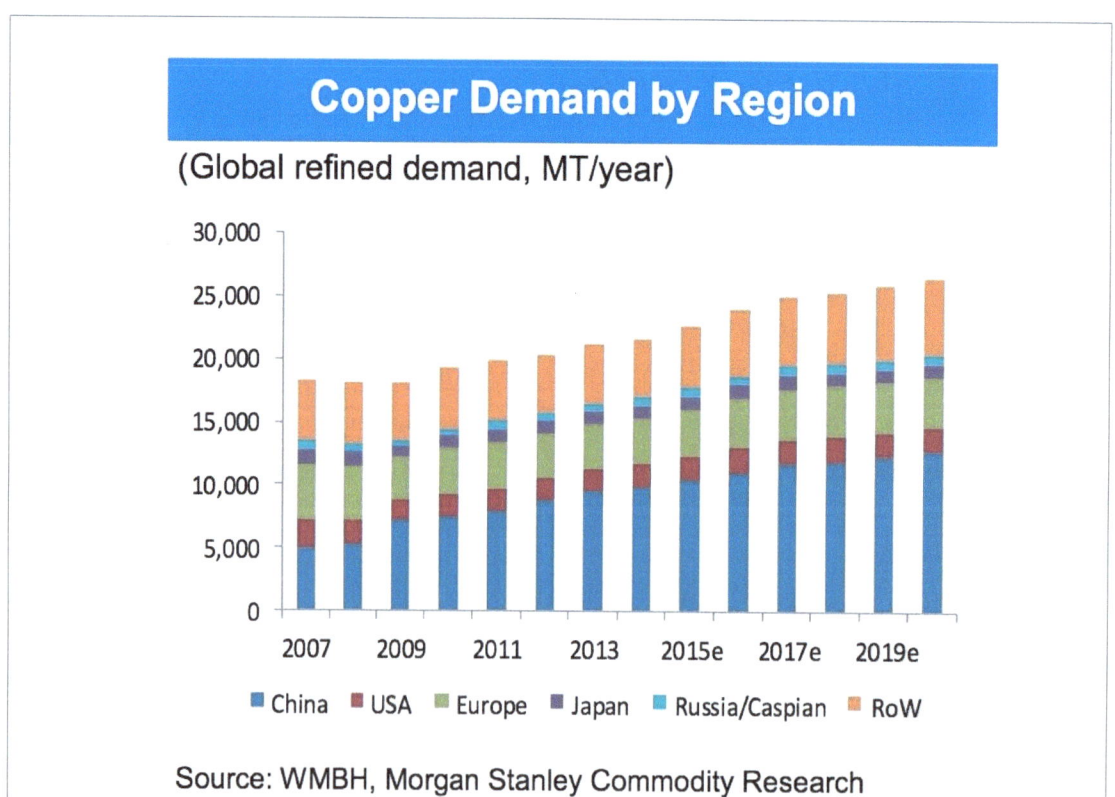

15.1 Using the Periodic Table

Summarize main points from each video.

Video Title / topic

Video Title / topic

Video Title / topic

Topic Introduction

Summarize your understanding of each paragraph.

Like sports, the study of atomic and molecular properties, and how they relate on the periodic table, is like going to practice. Learning about chemical reactions, is like stepping out onto the field for the game itself.

Just as every sport has its "vocabulary"—the concepts of offense and defense, as well as various rules and strategies—the study of chemical reactions involves a large set of terms.

If liquid water is boiled, it is still water; likewise frozen water, or ice, is still water. Melting, boiling, or freezing simply by the application of a change in temperature are examples of physical changes – not chemical changes.

A chemical change occurs when the actual composition changes— that is, when one substance is transformed into another. Water can be chemically changed, for instance, when an electric current is run through a sample, separating it into oxygen and hydrogen gas.

Read/Summarize Text

1. Read the passage.
2. Underline key expressions in each sentence.
3. Re-write each word (or expression) you underlined.
4. Summarize the passage.

Title of Passage.

> The relative position of an element on the periodic table is directly related to the likelihood of a reaction happening and even what kind of compound will be formed from a combination of elements.
>
> Central to a discussion of element reactivity is the concept of electronegativity. Electronegativity is a measure of the strength of electrical attraction between an element's nucleus and its valence (highest energy) electrons. Metals tend to be more reactive the further down the periodic table we look. This is especially true of the Alkali and Alkaline Earth Metals. Metal elements will likely lose electrons when they react.

http://learningchemistryeasily.blogspot.com

Re-write words you underlined

_____ _____ _____

_____ _____ _____

Using a complete sentence, summarize or rephrase the passage

Read Text for Comprehension

Read this article for deeper understanding. No summary is required, although you may want to circle, underline, or mark key ideas and words.

https://www.pinterest.com/explore/chemical-reactions/?lp=true

Household items can be useful learning tools for chemical reactions. Some of these can be dangerous – so BEFORE conducting a "household item" chemical combination, research the combination through the internet and other sources.

Commonly used materials from the kitchen and household supply of items:

Baking Soda

Vinegar

Water

Soda (Coke)

Lemon / Lemon Juice

Potato

Hydrogen Peroxide

Sodium Iodide

Table Salt

Sugar

Balloons

Bottles

Pennies

(Old) dimes

Hair follicles

Milk

Draw Illustration

Copy and Label the Illustration in the Space Provided

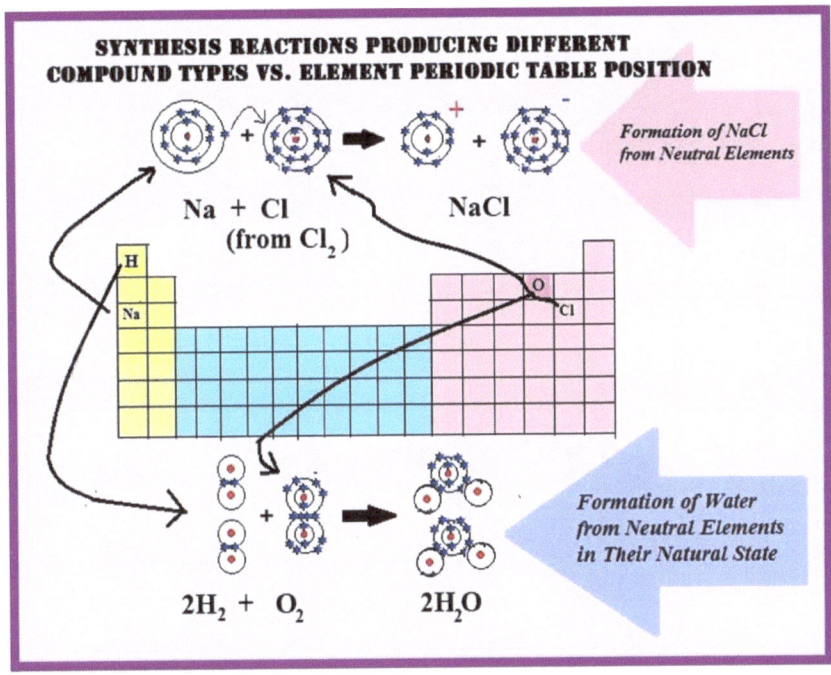

http://learningchemistryeasily.blogspot.com/2013/04/chemical-reactions-and-periodic-table.html

Draw (Copy) the Illustration Here

Interpret a Graph

Write the title of the graph _____

Circle the type of chart this represents
       Bar Chart   Line Chart   Pie Chart   Other

If applicable,
    What does the X-axis represent _____

    What does the Y-axis imply _____

Summarize what this graph represents or conveys

_____

https://socratic.org/questions/how-can-chemical-reactions-affect-the-decomposition-of-important-nutrients-in-fo

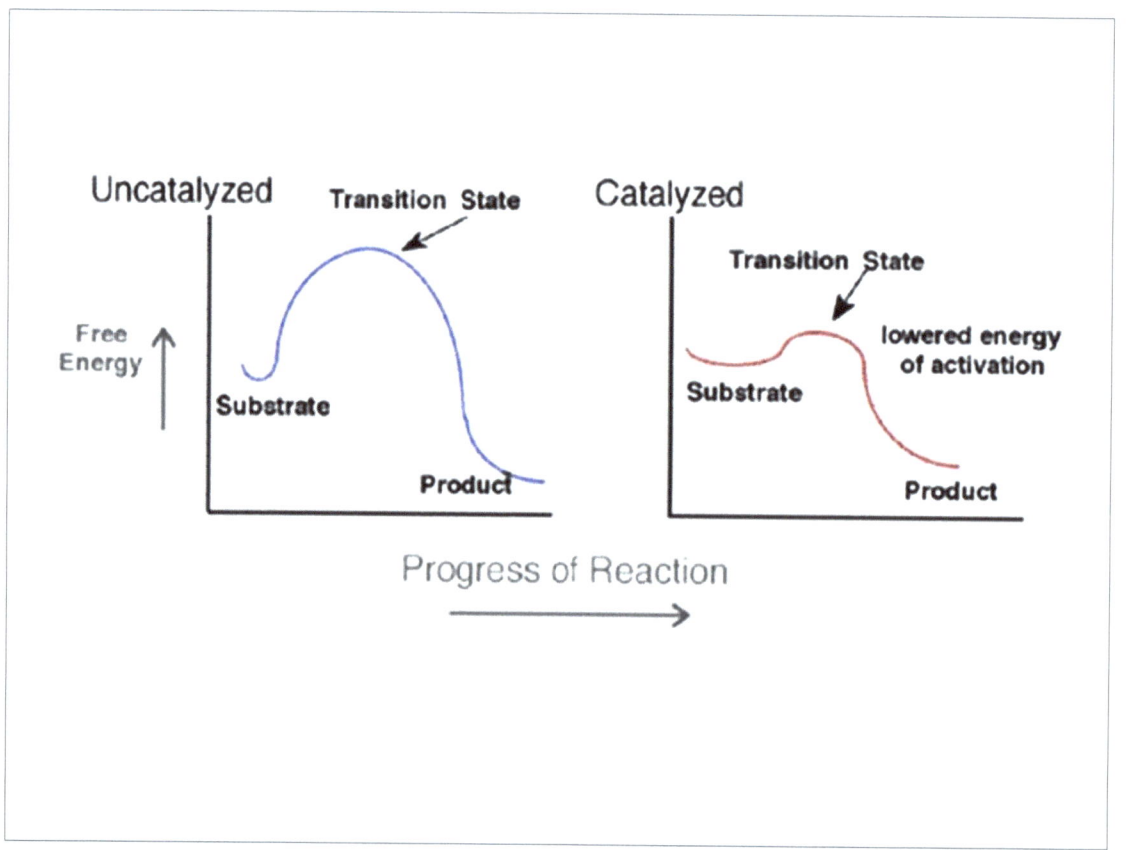

16.1 Naming Binary Compounds

Summarize main points from each video.

Video Title / topic

Video Title / topic

Video Title / topic

Topic Introduction

Summarize your understanding of each paragraph.

In the earliest years of chemistry as a science, scientists had not yet settled on a consistent way to name elements and compounds. Many of the earliest names used are still used today. These are referred to as a "common name."

In an ideal world, a system for naming would exist so that anyone familiar with the naming system could identify an element, a compound, or a formula. While the common names still are used, today there are several "rules" for naming compounds.

In addition to the "rules" for naming a compound – chemistry students must also memorize several names that do not follow those rules. The names for compounds that "break the rules" are often some of the most frequently cited compounds.

One straight-forward rule for naming "Type I" ionic compounds is to list the cation first – and the anion second. A good example of this is table salt. Table salt has the chemical formula of NaCl. The two ions for the compound are Na^+ and Cl^-. The name is sodium chloride.

Read/Summarize Text

1. Read the passage.
2. Underline key expressions in each sentence.
3. Re-write each word (or expression) you underlined.
4. Summarize the passage.

Rules for Naming Binary Ionic Compounds

1. The full name of the cation is listed first.

2. The root of the anion name is listed second and is followed by the suffix "ide."

3. If the compound contains a transition metal, a Roman numeral is included after the metal name to indicate the oxidation number of the metal.

Examples
$NaCl$ – sodium chloride
BaF_2 – barium fluoride
NH_4OH – ammonium hydroxide

www.stetson.edu/~wgrubbs/genchem1/namingcompounds

Re-write words you underlined

_____   _____   _____

_____   _____   _____

Using a complete sentence, summarize or rephrase the passage

33

Read Text for Comprehension

Read this article for deeper understanding. No summary is required, although you may want to circle, underline, or mark key ideas and words.

Binary compounds are compounds that consist of two elements. There are three types of binary compounds. Binary compounds containing:

- two nonmetals (eg - CO)
- metals with fixed ionic charges (eg - AgCl)
- metals with variable ionic charges (eg - FeS)

In naming, each type of binary compound follows its set of rules.

Binary compounds containing two nonmetals

Rules for naming binary compounds containing two nonmetals:

1. Name the first element by its name.
2. The second element has the ending -ide.
3. The number of atoms of each element is indicated with Greek prefixes. In the case of mono-, it is only used for the second nonmetal. When no prefix appears, one atom is assumed.
4. If two vowels appear next to each other, the vowel from the Greek prefix is dropped. This is for ease of pronunciation.
    - monooxide becomes monoxide
    - tetraoxide becomes tetroxide
    - pentaoxide becomes pentoxide

Examples

CO -	carbon monoxide
CO2 -	carbon dioxide
CCl4 -	carbon tetrachloride
SO2 -	sulfur dioxide
N2O4 -	dinitrogen tetroxide

http://nobel.scas.bcit.ca/wiki/index.php/Naming_binary_compounds

Draw Illustration

Copy and Label the Illustration in the Space Provided

Illustration

Cations		Anions	
Name	Formula	Name	Formula
Hydrogen	H^+	Hydroxide	OH^-
Sodium	Na^+	Chloride	Cl^-
Potassium	K^+	Nitrate	NO_3^-
Ammonium	NH_4^+	Acetate	CH_3COO^-
Silver	Ag^+	Bicarbonate	HCO_3^-
Calcium	Ca^{2+}	Sulfide	S^{2-}
Iron(II)	Fe^{2+}	Oxide	O^{2-}
Copper	Cu^{2+}	Sulfate	SO_4^{2-}
Lead	Pb^{2+}	Carbonate	CO_3^{2-}
Iron(III)	Fe^{3+}		
Aluminum	Al^{3+}		

https://sites.google.com/site/xyuhanliu99x/letter/readingassignment5textbookchapter6ionicbonding

Draw (Copy) the Illustration Here

Interpret a Graph

Write the title of the graph _____

Circle the type of chart this represents

    Bar Chart   Line Chart   Pie Chart   Other

If applicable,

    What does the X-axis represent _____

    What does the Y-axis imply _____

Summarize what this graph represents or conveys

_____

https://mcgroup.co.uk/researches/caustic-soda

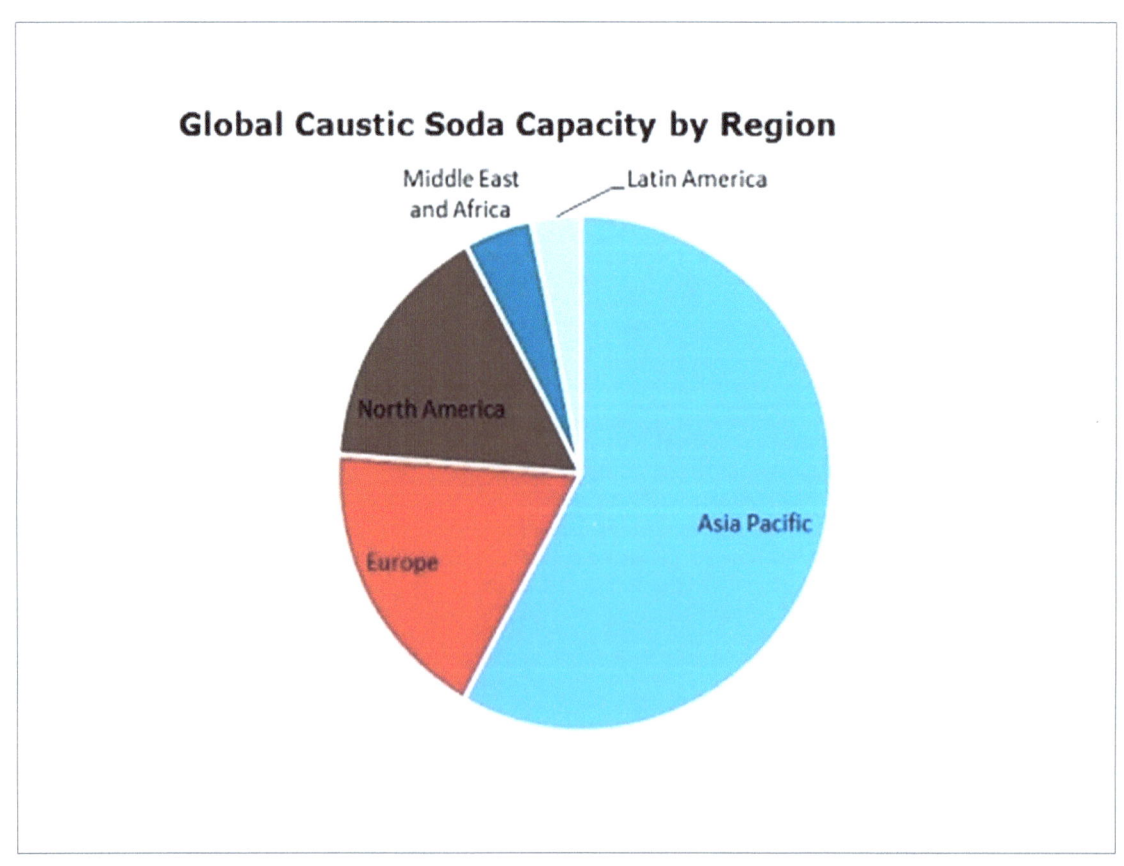

17.1 Chemical Reactions

Summarize main points from each video.

Video Title / topic _____

Video Title / topic _____

Video Title / topic _____

Topic Introduction

Summarize your understanding of each paragraph.

Chemical reactions occur when substances go through chemical change. The substances go through change resulting in new substances. Chemical reactions are not the same thing as a physical change. Recall that physical changes are a change in state of matter.

Chemical reactions rearrange atoms. Note that energy is conserved in chemical reactions. Even though energy may APPEAR to not be conserved – the total energy of everything combined remains the same amount of energy before and after the reaction.

Chemical reactions are sometimes exothermic. Sometimes they are endothermic. You need to remember this fact, and to distinguish between the two words.

Exothermic means that energy is released to the surroundings.

Endothermic reactions need more energy to break the bonds in the reactants than is given off by forming bonds in the products.

Read/Summarize Text

1. Read the passage.
2. Underline key expressions in each sentence.
3. Re-write each word (or expression) you underlined.
4. Summarize the passage.

Describing Reactions

You can describe the reaction between atoms and molecules in several ways. One way is to write a word equation. A word equation shows the names of the products and reactants. Another way is to use molecular models, which can be used to show how the atoms are rearranged during the reaction.

The clearest way is to write a chemical equation. A chemical equation uses symbols (from the Periodic Table) to represent a chemical reaction. The equation shows the relationship between the reactants and the products of a reaction. Here is an example:

$$CH_4 + 2O_2 \Rightarrow CO_2 + 2H_2O$$

Adapted from Physical Science (Holt) Chemical Equations page 225.

Re-write words you underlined

_____   _____   _____

_____   _____   _____

Using a complete sentence, summarize or rephrase the passage

Read Text for Comprehension

Read this article for deeper understanding. No summary is required, although you may want to circle, underline, or mark key ideas and words.

Stoichiometry

Pretend you want to make chocolate chip cookies. You have a great recipe handed down from your grandmother that calls for two cups of chocolate chips, but you only have one cup of chocolate chips in the house. It's raining outside, and you don't feel like going to the store. So, what do you do? Do you make the cookies with half the chocolate chips the recipe calls for? No way! Who wants to eat cookies with only half the chocolate?

Instead, you determine the ratio of chocolate chips on hand to amount needed, which is 1:2. Then, you adjust the ratio of all the other ingredients in the recipe. Essentially, you have just performed stoichiometry, one of the fundamental aspects of chemistry. Stoichiometry is a word derived from two Greek words: 'stoicheon' meaning element, and 'metron,' meaning measure. This is pretty cool because stoichiometry is essentially the measurement of elements, or the study of chemical quantities consumed or produced in a chemical reaction.

Performing stoichiometry involves the use of a special chemical counting unit: the mole. Just to review for a moment, a mole isn't an animal. Well, it is, but not in chemistry. In chemistry, a mole is a unit of measurement, such that one mole of a substance contains $6.022*10^{23}$ particles.

In chemistry, particles can be atoms, molecules, or compounds. Conveniently, one mole of a substance has a mass that is equal to its atomic mass expressed in grams. This relationship is known as molar mass. For example, one atom of carbon has a mass of 12.011 amu, one mole of carbon has a mass of 12.011 grams.

When we do stoichiometry, we always want to speak about chemicals in terms of how many moles are present. The essence of stoichiometry involves comparing how many moles of chemicals are present. We may be simply comparing the number of moles of each reactant needed, or the number moles reactant to number of moles product.

http://study.com/academy/lesson/mole-to-mole-ratios-and-calculations-of-a-chemical-equation.html

Draw Illustration

Copy and Label the Illustration in the Space Provided

Reaction Types

Combination
Decomposition
Substitution
Double-substitution
Combustion

HoneycuttScience.com

Draw (Copy) the Illustration Here

Answer the Three Questions

Subscripts vs. Coefficients

3 CuCl$_2$ + 2 Al → 2 AlCl$_3$ + 3 Cu

On the *reactant side* of the equation:

How many **copper (II) chloride molecules** react?

How many **total chlorine atoms** are there?

How many **aluminum atoms** are there?

http://slideplayer.com/slide/9189050/

18.1 Scientific Notation and Units

Summarize main points from each video.

Video Title / topic _____

Video Title / topic _____

Video Title / topic _____

Topic Introduction

Summarize your understanding of each paragraph.

Scientific notation is a way of expressing numbers that are too big or too small to be conveniently written in decimal form. It is commonly used by scientists, mathematicians and engineers, in part because it can simplify certain arithmetic operations.

Normalized Notation. Any given integer can be written in the form $m \times 10^n$ in many ways: for example, 350 can be written as 3.5×10^2 or 35×10^1 or 350×10^0.

Units of Measure. The SI base units and their physical quantities are the meter for length, the kilogram for mass, the second for time, the ampere for electric current, the Kelvin for temperature, the candela for luminous intensity, and the mole for amount of substance.

Best Estimate ± Uncertainty. When scientists make a measurement or calculate some quantity from their data, they generally assume that some exact or "true value" exists based on how they define what is being measured (or calculated).

Wikipedia.com

Read/Summarize Text

1. **Read the passage.**
2. **Underline key expressions in each sentence.**
3. **Re-write each word (or expression) you underlined.**
4. **Summarize the passage.**

Title of Passage.

> A metric prefix is a unit prefix that precedes a basic unit of measure to indicate a multiple or fraction of the unit. While all metric prefixes in common use today are decadic, historically there have been a number of binary metric prefixes as well.
>
> Each prefix has a unique symbol that is prepended to the unit symbol. The prefix kilo-, for example, may be added to gram to indicate multiplication by one thousand: one kilogram is equal to one thousand grams. The prefix milli-, likewise, may be added to metre to indicate division by one thousand; one millimetre is equal to one thousandth of a metre.

https://en.wikipedia.org/wiki/Metric_prefix

Re-write words you underlined

_____   _____   _____

_____   _____   _____

Using a complete sentence, summarize or rephrase the passage

Read Text for Comprehension

Read this article for deeper understanding. No summary is required, although you may want to circle, underline, or mark key ideas and words.

Numbers in Science: Scientific Notation

When scientists measure a quantity, they actually measure two pieces of information--the value they think they have measured, and the uncertainty. This can be stated as "We measured ten, plus or minus one", and often scientists do use these terms. However, this notation gets cumbersome fast. We need a quick, generally accepted method by which we can indicate the precision of our measurements.

Scientists put only the digits they can reasonably be certain of in their numbers. They might say, for example, that they measured "10." cm (note the presence of the decimal point). This is actually different from saying that they measured "10" cm. The use of the decimal point indicates that the scientist is sure of both digits to some reasonable degree -- it is "10 point something", not 11 or 9, even though rounding both of these numbers to one digit gives 10.

The number "10." is said to have two significant digits, or significant figures, the 1 and the 0. The number 1.0 also has two significant digits. So does the number 130, but 10 and 100 only have one "sig fig" as written. Zeros that only hold places are not considered to be significant.

So, how does a scientist indicate that two of the digits in 100 are significant?? We can't put in a decimal point alone to make 100. because that would indicate 3 digits. What should we do?

Scientists use scientific notation to handle this problem. Scientific notation makes sure that everything but the first digit of a number is after the decimal place and therefore either certain or not used. Here are some numbers in scientific notation to study:

https://www.shodor.org/unchem/math/science/index.html

Draw Illustration

Copy and Label the Illustration in the Space Provided

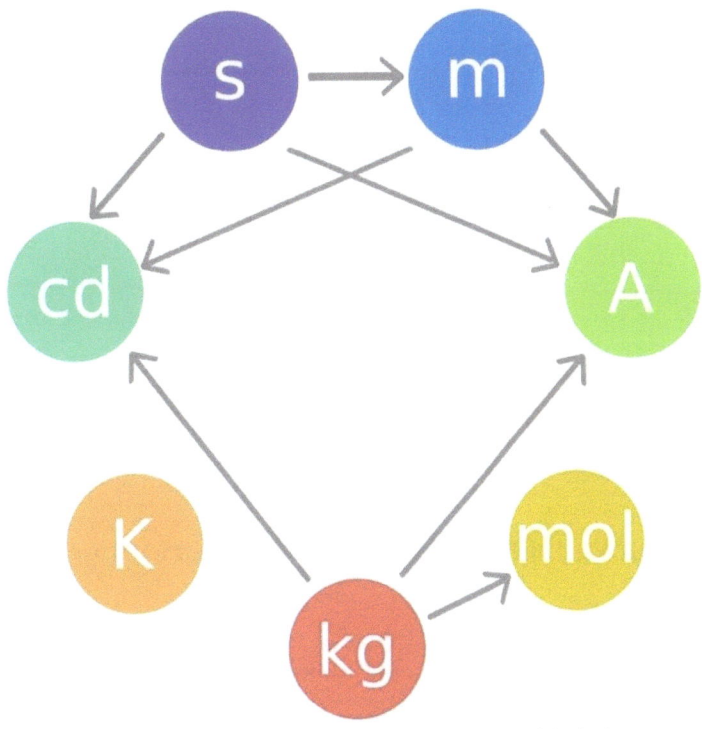

https://upload.wikimedia.org/wikipedia/commons/thumb/c/c8/SI_base_unit.svg/1200px-SI_base_unit.svg.png

Draw (Copy) the Illustration Here

Interpret a Graph

Write the title of the graph _____

Circle the type of chart this represents
      Bar Chart   Line Chart   Pie Chart   Other

If applicable,
    What does the X-axis represent _____

    What does the Y-axis imply _____

Summarize what this graph represents or conveys

_____

www.zmescience.com

United States   The Rest of the World

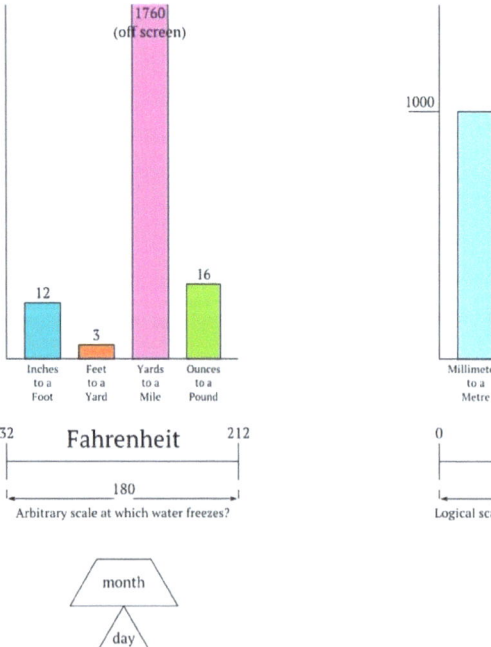

19.1 Atoms and Moles

Summarize main points from each video.

Video Title / topic _____

Video Title / topic _____

Video Title / topic _____

Topic Introduction

Summarize your understanding of each paragraph.

Most of the mass of atoms comes from the protons and the neutrons. The number of protons DEFINES the element. The number of neutrons often equals the number of protons. BUT the number of neutrons in an atom might be slightly fewer or more than the number of protons.

If the number of neutrons were ALWAYS equal to the number of protons, then the average mass of an atom would be double the atomic number. But the number of neutrons is not always equal to the number of protons. (Recall these are called isotopes).

The average mass of an element is expressed in grams. A sample of an element with a mass equal to that element's average atomic mass contains 1 mol of atoms. For example, 1 mol of Aluminum (1 mol Al) has 6.023×10^{23} atoms with a mass of 26.98 grams.

One mole of something consists of 6.023×10^{23} units of that substance. That's a huge number, and somewhat difficult to imagine. A common way to visualize that big number is that one mole of marbles would cover the earth to a depth of 50 miles!

Read/Summarize Text

1. Read the passage.
2. Underline key expressions in each sentence.
3. Re-write each word (or expression) you underlined.
4. Summarize the passage.

Robert Millikan's Contribution to Avogadro's number

Accurate determinations of Avogadro's number require the measurement of a single quantity on both the atomic and macroscopic scales using the same unit of measurement. This became possible for the first time when American physicist Robert Millikan measured the charge on an electron.

Millikan made numerous momentous discoveries in the fields of electricity, optics, and molecular physics. His earliest major success was the accurate determination of the charge carried by an electron, using the elegant "falling-drop method"; he also proved that this quantity was a constant for all electrons (1910), thus demonstrating the atomic structure of electricity.

www.nobelprize.org
www.scientificamerican.com

Re-write words you underlined

_____   _____   _____

_____   _____   _____

Using a complete sentence, summarize or rephrase the passage

Read Text for Comprehension

Read this article for deeper understanding. No summary is required, although you may want to circle, underline, or mark key ideas and words.

Avogadro's number is a dimensionless quantity, and has the same numerical value of the Avogadro constant when given in base units.

The Avogadro constant is named after the early 19th-century Italian scientist Amedeo Avogadro, who, in 1811, first proposed that the volume of a gas (at a given pressure and temperature) is proportional to the number of atoms or molecules regardless of the nature of the gas.

The French physicist Jean Perrin in 1909 proposed naming the constant in honor of Avogadro. Perrin won the 1926 Nobel Prize in Physics, largely for his work in determining the Avogadro constant by several different methods

Accurate determinations of the Avogadro constant require the measurement of a single quantity on both the atomic and macroscopic scales using the same unit of measurement. This became possible for the first time when American physicist Robert Millikan measured the charge on an electron in 1910. The electric charge per mole of electrons is a constant called the Faraday constant and had been known since 1834 when Michael Faraday published his works on electrolysis. By dividing the charge on a mole of electrons by the charge on a single electron the value of Avogadro's number is obtained.

The Avogadro constant is a scaling factor between macroscopic and microscopic (atomic scale) observations of nature.

As may be observed in the table below, the main limiting factor in the precision of the Avogadro constant is the uncertainty in the value of the Planck constant, as all the other constants that contribute to the calculation are known more precisely.

Constant	Symbol	2014 CODATA value	Relative standard uncertainty	Correlation coefficient with N_A
Proton-electron mass ratio	m_p/m_e	1836.152 673 89(17)	9.5×10^{-11}	−0.0003
Molar mass constant	M_u	0.001 kg/mol = 1 g/mol	0 (defined)	—
Rydberg constant	R_∞	10 973 731.568 508(65) m^{-1}	5.9×10^{-12}	−0.0002
Planck constant	h	6.626 070 040(81) × 10^{-34} J s	1.2×10^{-8}	−0.9993
Speed of light	c	299 792 458 m/s	0 (defined)	—
Fine structure constant	α	7.297 352 5664(17) × 10^{-3}	2.3×10^{-10}	0.0193
Avogadro constant	N_A	6.022 140 857(74) × 10^{23} mol^{-1}	1.2×10^{-8}	1

https://en.wikipedia.org/wiki/Avogadro_constant

Fill-in the Missing Information

Atomic Number	Element	Symbol	Atomic Mass
1	_____	H	1.0079
2	Helium	He	_____
___	_____	Li	_____
4	Beryllium	Be	_____
5	_____	B	10.8110
6	_____	C	_____
___	Nitrogen	N	_____
___	_____	O	15.9994
___	Fluorine	F	18.9984
10	_____	Ne	20.1797
11	_____	Na	22.9897
12	_____	Mg	24.3050
13	_____	Al	26.9815
___	Silicon	Si	_____
___	Phosphorus	P	_____
___	Sulfur	S	_____
___	Chlorine	Cl	35.4530
___	Potassium	K	_____
___	_____	Ar	39.9480
20	_____	Ca	40.0780
21	Scandium	Sc	_____
___	Titanium	Ti	47.8670
___	Vanadium	V	50.9415
___	Chromium	Cr	_____
25	_____	Mn	54.9380

Draw Illustration

Copy and Label the Illustration in the Space Provided

Illustration

DIVIDE → DIVIDE →
Atoms or Molecules — Mols — Grams
← MULTIPLY ← MULTIPLY

Convert Using 6.02×10^{23}

Convert Using Atomic/Molar Mass

www.boundless.com

Draw (Copy) the Illustration Here

23.1 Reactions in Aqueous Solutions

Summarize main points from each video.

Video Title / topic

Video Title / topic

Video Title / topic

Topic Introduction

Summarize your understanding of each paragraph.

An aqueous solution is a solution in which the solvent is water. It is usually shown in chemical equations by appending (aq) to the relevant chemical formula. For example, a solution of table salt, or sodium chloride (NaCl), in water would be represented as Na+(aq) + Cl−(aq).

The word aqueous means pertaining to, related to, similar to, or dissolved in water. As water is an excellent solvent and is also naturally abundant, it is a ubiquitous solvent in chemistry.

Substances that are hydrophobic ('water-fearing') often do not dissolve well in water, whereas those that are hydrophilic ('water-friendly') do. An example of a hydrophilic substance is sodium chloride.

Reactions in aqueous solutions are usually metathesis reactions. Metathesis reactions are another term for double-displacement; that is, when a cation displaces to form an ionic bond with the other anion.

https://en.wikipedia.org/wiki/Aqueous_solution.

Read/Summarize Text

1. Read the passage.
2. Underline key expressions in each sentence.
3. Re-write each word (or expression) you underlined.
4. Summarize the passage.

Solubility

Solubility is the property of a solid, liquid, or gaseous chemical substance called solute to dissolve in a solid, liquid, or gaseous solvent. The solubility of a substance fundamentally depends on the physical and chemical properties of the solute and solvent as well as on temperature, pressure and the pH of the solution. The extent of the solubility of a substance in a specific solvent is measured as the saturation concentration, where adding more solute does not increase the concentration of the solution and begins to precipitate the excess amount of solute. The solubility of a substance is an entirely different property from the rate of solution, which is how fast it dissolves.

https://en.wikipedia.org/wiki/Solubility

Re-write words you underlined

_____ _____ _____

_____ _____ _____

Using a complete sentence, summarize or rephrase the passage

Read Text for Comprehension

Read this article for deeper understanding. No summary is required, although you may want to circle, underline, or mark key ideas and words.

Example (1) - Let us consider the possible reaction of aqueous solution of NaCl with aqueous solution of $AgNO_3$. We would place a few drops of the NaCl solution in the reaction container followed by a few drops of $AgNO_3$ solution and observe an immediate cloudiness (white precipitate) that indicates a solid precipitate has formed. A precipitation chemical reaction has occurred.

In order to determine the possible identity of the solid product that forms, we first identify the ions present in each of the two aqueous solutions we started with: Na^+, Cl^- (from NaCl) and Ag^+, NO_3^- (from $AgNO_3$).

Next we examine the ions for possible new combinations that may lead to a reasonable product formula. The combination of ions (NaCl, $AgNO_3$) that existed in solution prior to the experiment had been soluble and therefore should remain as such without separating out as solid after the reaction. This allows us to eliminate combinations like NaCl and AgNO3 from the list of possibilities.

This leaves us with only two other possibilities, AgCl and $NaNO_3$. From the knowledge of **Solubility Rules** we can determine which of these two products is insoluble. Solubility Rule indicates that nitrate salts are soluble. Therefore, $NaNO_3$ cannot be the precipitate in this reaction. Also solubility Rule states that most chloride salts are soluble. AgCl is listed as an exception to this rule. In this case, *it is* AgCl which is the precipitate.

Once the chemical identity of the solid product is determined, we can then determine the **balanced formula equation**, the **complete ionic equation** as well as the **net ionic equation**, describing the chemistry that has occurred.

a) The **balanced formula equation** for the reaction of aqueous $AgNO_3$ with NaCl is written as:

$$AgNO_3(aq) + NaCl(aq) \rightarrow AgCl(s) + NaNO_3(aq)$$

Note that in the above equation, the physical state of the AgCl product is denoted by the letter s, to indicate that it is the precipitate. The number of atoms of each element is same before and after the reaction, indicating that the equation is balanced.

b) The **complete ionic equation**, indicates which reactants and products exist as ions and which ones do not:

$$Ag^+(aq) + NO_3^-(aq) + Na^+(aq) + Cl^-(aq) \rightarrow AgCl(s) + Na^+(aq) + NO_3^-(aq)$$

The ions that actually undergo change in the chemical reaction and participate in the formation of the insoluble product are called ***participating ions***. In the above reaction, Ag^+ and Cl^- are the participating ions. Those that do not undergo change are called ***spectator ions***. In the above reaction, Na^+ and NO_3^- are the spectator ions.

c) The **net ionic equation** displays only the participating ions on the reactant side, and the precipitate on the product side. The physical states of the reactants and products are also indicated. The spectator ions are not included.

$$Ag^+(aq) + Cl^-(aq) \rightarrow AgCl(s)$$

http://swc2.hccs.edu/pahlavan/intro_labs/Exp_10_Precipitation_Reactions_(Metathesis_Reactions).pdf

Draw Illustration

Copy and Label the Illustration in the Space Provided

Types of Stoichiometry

Types	Explanation
Mole- Mole	We have to relate the moles of the reactants with moles of the product
Mole- Mass	We have to relate the moles of the reactants with the mass of the products
Mass- Mole	We have to relate the mass of the reactants with the moles of the product
Mass- Mass	We have to relate the mass of the reactants with the mass of the products

https://www.slideshare.net/pritinayak/stoichiometry-part-1-introduction

Draw (Copy) the Illustration Here

Interpret a Graph

Write the title of the graph _____

Circle the type of chart this represents
      Bar Chart   Line Chart   Pie Chart   Other

If applicable,
    What does the X-axis represent _____

    What does the Y-axis imply _____

Summarize what this graph represents or conveys

_____

https://www.engineeringtoolbox.com/density-aqueous-solution-inorganic-sodium-salt-concentration-d_1957.html

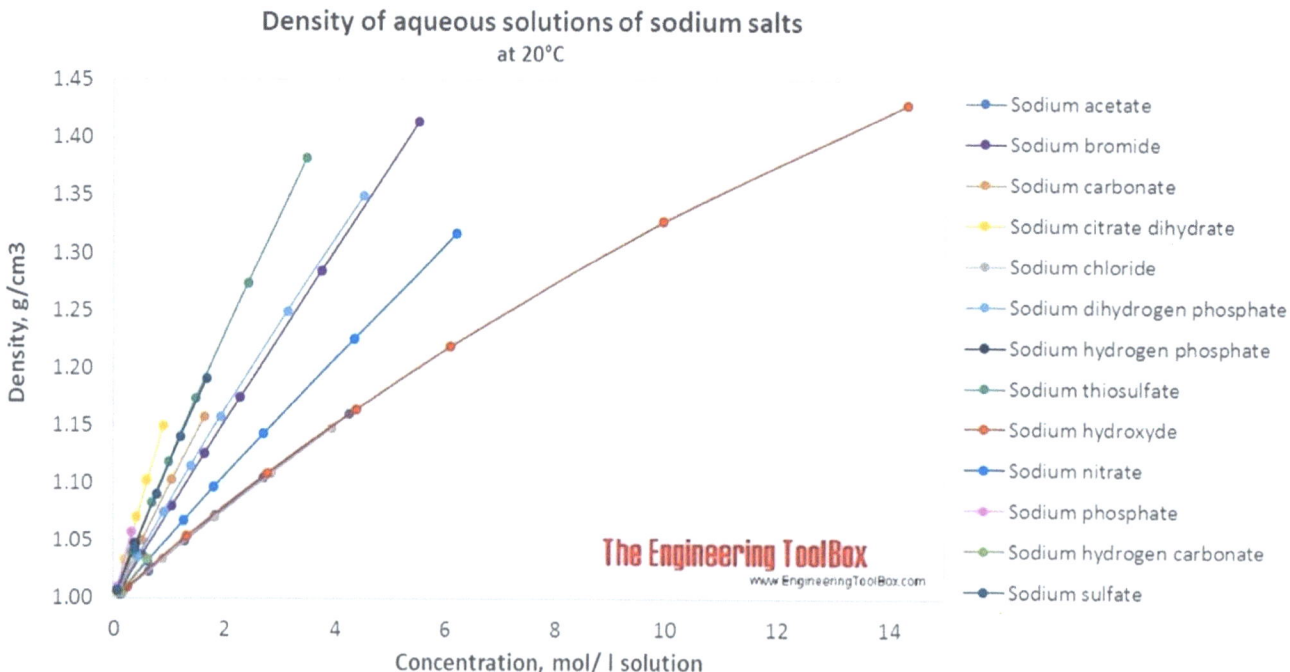

GLOSSARY - VOCABULARY

Allotropes
Different forms of the same element

Atom
Element composed of one type, the {blank}.

Atomic Number
Number of protons in an atom is equal to a number called {blank}.

Average Atomic Mass
Is the weighted average mass of its isotopes.

Binary Compound
Composed of two elements.

Boiling Point
The point in temperature when the liquid starts to boil.

Buoyancy
Ability of a fluid to exert an upward force on an object immersed in it.

Chemical Bond
Force that holds atoms together in a compound.

Chemical Change
A change of one substance to another.

Chemical Formula
A formula that shows what elements are in a compound and what it will become.

Chemical Property
Characteristic of a substance that indicates whether it can undergo a chemical change.

Chemical Reaction
More than one substances turning into other substances.

Coefficients
Numbers in front of each substance in a equation.

Colloid
Mixture with particles larger than those in solutions, but not heavy enough to settle out.

Combustion Reaction
Substance reacts with oxygen to make heat and light.

Compound
A substance in which the atoms of 2 or more elements are combined.

Covalent Bond
Attraction formed when elements share electrons.

Decomposition Reaction
One substance breaks down, into 2 more.

Diatomic Molecule
Consists of 2 atoms of the same element in a covalent compound.

Diffusion
Spreading of particles throughout a given volume until they are distributed.

Distillation
A process for separating substances by evaporating a liquid and recondensing its vapor.

Double Displacement Reaction
Two elements replace another to make a product.

Ductile
Flexible.

Electron Cloud
Area around a nucleus where electrons are mostly found.

Electron Dot Diagram
Uses the symbol of the element and dots to represent electrons.

Electrons
Particles in a atom with a negative charge.

Groups
   The vertical columns in the periodic table.

Heat of Fusion
   The energy required to change a substance from a solid to a liquid at its melting point.

Heat of Vaporization
   Amount of energy required for a liquid to become a gas.

Heterogeneous Mixture
   Mixture in which different materials can easily be distinguished.

Homogeneous Mixture
   Contains 2 or more gases, liquids, or solids substances blended evenly.

Ionic Bond
   Force of attraction between opposite charges.

Isotopes
   Atoms of the same element that have different numbers of neutrons.

Kinetic Theory
   Explanation of how particles in matter behave.

Law of Conservation and Mass
   The mass before a chemical change equals the mass of all the substances after the change.

Malleable
   Can be hammered.

Mass Number
   This number is the sum of the number of protons and neutrons.

Melting Point
   The point in temperature when the solid starts to liquefy.

Metallic Bonding

Metals
Good conductors of heat and electricity.

Molecule
A neutral molecule that forms as a result of electron sharing.

Neutrons
Particles in a atom with no charge.

Nonmetals
Usually gasses or brittle solids at room temperature.

Nucleus
The center of a atom.

Oxidation Number
Tells you how many electrons an atom has gained or lost.

Pascal
Used to measure pressure.

Periodic Table
A table filled with elements in order of atomic numbers, etc.

Periods
Horizontal rows of elements.

Physical Change
A change in size, shape, or state.

Physical Property
Characteristic of a material you can observe without changing the identity.

Polyatomic Ion
Positively or negatively charged, covalently bonded group.

Pressure
Force exerted per area. (Formula)

Products
Substances that are made.

Protons
    Particles in a atom with a positive charge.

Quarks
    Smaller particles in neutrons and protons.

Reactants
    Substances that react.

Semiconductors
    Elements that conduct under circumstances.

Single Displacement Reaction
    One element replaces another to make a product.

Solution
    Homogeneous mixture with particles so small that they cannot be seen with a microscope.

Sublimation
    The process of a solid going directly into a vapor.

Substance
    A type of matter with a fixed composition.

Suspension
    Heterogeneous mixture containing a liquid where visible particles settle.

Synthesis Reaction
    2 or more substances that combine to make another.

Transitional Elements
    Between groups 1 and 2, and 13 and 18.

Tyndall Effect
    Scattering of light by colloidal particles.

Viscosity
    The resistance to flow by a fluid.

White Space for Notes

References

Web Sites

acialloys.com
bcit.ca
boundless.com
dictionary.com
ency123.com
google.com/site
hccs.edu
lumenlearning.com
madsci.com
mcgroup.co.uk
nobleprize.org
answers.yahoo.com
businessinsider.com
compoundchem.com
engineeringtoolbox.com
honeycuttscience.com
learningchemistryeasy
scientificamerican.com
npl.co.uk
pinterest.com
sciencelearn.org
shodor.org
soctratic.org
stetson.edu
study.com
wikibooks.org
wikimedia.org
wikipedia.org
zmescience.com

Text Books

Allison, M. A. (2010). Austin, TX: Holt McDougal.

Zumdahl, S. S. (2007). Belmont, CA: Brooks/Cole.

Nowiki, S. (2012). Orlando, FL: Houghton Mifflin Harcourt Publishing Company.

Dobson, K. (2008). Austin, TX: Holt, Rinehart and Winston.

For more information, visit

www.HoneycuttScience.com